Paulo Jean Pereira Alves

Urban solid waste: a source of sustainability

Paulo Jean Pereira Alves

Urban solid waste: a source of sustainability

Work and sustainability at the dump in the city of Pires do Rio -GO

ScienciaScripts

Imprint

Cover image: www.ingimage.com

This book is a translation from the original published under ISBN 978-613-9-67375-9.

Publisher:
Sciencia Scripts
is a trademark of
Dodo Books Indian Ocean Ltd. and OmniScriptum S.R.L publishing group

120 High Road, East Finchley, London, N2 9ED, United Kingdom
Str. Armeneasca 28/1, office 1, Chisinau MD-2012, Republic of Moldova, Europe
Printed at: see last page
ISBN: 978-620-8-15983-2

SUMMARY

I dedicate this research to my mother, Marleusa Pereira Alves, who is largely responsible for who I am today.

Thanks

To God first, for He is worthy of all honor, glory and praise, ES the author of life.

To my mother Marleusa P. Alves, who supports me so much in my decisions.

To my grandmother, Jùlia, who, even at her advanced age, vibrates and cheers on my achievements.

To my siblings, nephews, cousins and all my family who are always by my side.

To Professor Ms. Cristiane Dias for her dedication and friendship.

To Professor MS. Marcelo Venâncio, who, even with a tight schedule, always attended to his students with respect and dedication.

To the waste pickers ("Garbage Professionals"), humble and dignified people who play an extremely important role in the environmental context, even though they and society in general are not aware of this fact.

To every colleague in the Geography class, for the moments we spent together over these four years, even those who were transferred, or those who could not complete the course due to force majeure, all of whom I will always remember. In particular, I would like to thank my great friend Fernando "o gala do celular", my great and eternal friend Luciana Ananias "bota", my companion Leonardo "cabeça gràvida", who always had time to exchange ideas, and Wellivan "gordim", who was always there for all four years.

To the professors, who each year, with dedication, knew how to transfer their knowledge to us.

In a special way, I would like to thank my close friend and teacher Ms. Maria Eni S. D. Freire for all her effort and commitment over the last four years.

BEATING THE OME

We're tired of this life of bitterness

Old iron and city hall trying to dominate us

Carrying a hundred kilo bale on his back

And the doctor with his pen and more laws to screw us over

The waste picker knows that there is a movement

Which prepares the confrontation for popular power

Self-management of the production chain

Direct action every day to achieve socialism

Come scavenger

Pro organized movement

Because united we stand strong

And we will no longer be exploited

Come and recover your dignity

Fighting to build socialism with freedom!

I'm a waste picker and I'm always in a bind

When it rains I eat very little

And in the summer I only get sick

But I care about the town hall and the junkyard

They want to see us in a cemetery

Or in a hospital bed

I sort everything I find on my desk

I recycle nature with no boss to boss us around

Organized I temper rebellion

And when I see the bourgeoisie, I know it's i

(Erik Soares)

SUMMARY

Until the Industrial Revolution, the waste produced was organic, easily decomposed by nature, but with industry came new consumer habits. In the globalized world, new technologies have created wealth for some and poverty for others. In the midst of so much information and cutting-edge technologies, recyclable material collectors have emerged, and they are important players in sustainable development and the need for ergonomic intervention in the occupational activities of these individuals. In the city of Pires do Rio (GO), there are serious environmental problems resulting from a lack of urban planning. The huge increase in consumption generates enormous amounts of waste of all kinds, from homes, public activities and the industrial process. The waste produced in the city of Pires do Rio (GO) is collected by press trucks and dumped in an open dump, without any care being taken to separate dry and wet waste. The aim of this study is to understand how these people (waste pickers) live, why they are subjected to such a situation, to verify the perceptions of the residents living near the dump in relation to the waste pickers, and to analyze sustainability through waste collection. To this end, the research was carried out through bibliographical discussions, field research with nearby residents and an analysis of sustainability with the waste pickers through formal conversation and interviews. Several people collect and separate waste as a way of surviving, since society does not offer them other work opportunities. In the end, we presented the idea of organizing these waste pickers into cooperatives, as they could find a logical and innovative way out of the capitalist world, which is an alternative way of strengthening their work, always focused on improving their work.

Key words: Waste, waste pickers, sustainability and Pires do Rio.

INTRODUCTION

As with many other categories, waste has different connotations depending on the social and cultural situation in which people find themselves. However, in general it is possible to highlight that garbage is what is put "out of the house", what you don't want to come into contact with.

With globalization, consumption has increased, there has been technological evolution and the creation of means capable of generating comfort, thus affecting the environment, from fires, deforestation, air pollution caused by the emission of carbon dioxide from industries and garbage. Waste is just one of the problems that degrade the environment. Its nuisances are caused by the lack of awareness of the population, who throw dirt and waste in inappropriate places. Nature can no longer keep up with the production of garbage, which is why the damage it causes is increasing.

A dump is simply an open-air garbage dump, without any sanitary criteria to protect the environment, which allows full access to disease vectors such as flies, mosquitoes, cockroaches, rats and others. It is the place where urban and industrial waste accumulates in a rustic way, without any treatment. As there is no treatment for the waste deposited in the dump, it is mixed with household, hospital and industrial waste, causing various environmental problems, as well as the needy and unemployed population who are attracted to the dump, most of whom feed on the waste found there.

The constant precariousness of work, inherent in the capitalist production process, ends up highlighting the growing number of unemployed people plaguing the world.

The presence of several people in the Pires do Rio - GO dump was what drew us to the subject, because even with the risks to which they are exposed, such as a natural fire caused by the gases generated by the waste, they are managing to make a living and generate income by collecting and selling the waste.

Given this context, we raised the following problems:

- Could it be that garbage is a means of survival for some residents of Pires do Rio?
- What are the environmental consequences of the Pires do Rio dump?
- Is the way the garbage is collected in Pires do Rio suitable for the families who survive on it?

The aim is to analyze the social sustainability of the garbage dump in the city of Pires do Rio (GO) by showing how these people (waste pickers) live, why they are subjected to such a situation, also understanding the action of the social relations of public power, verifying the perception of the residents of the vicinity of the dump in relation to the way of working developed by the waste pickers and analyzing sustainability through waste collection.

To answer these problems and develop the proposed research, we used bibliographical research (or secondary sources), considering themes such as concepts and theories about garbage, dumps, landfills, sustainability, among others. To do this, we went into the field, dividing it into two stages, the first being systematic observation to verify and diagnose the place where solid urban waste is collected, as well as taking photographs. In the second stage, the sustainability of the waste pickers will be analyzed through formal conversations and interviews.

In fact, there is worldwide concern about the waste produced by society and we know that it is growing alarmingly. If we look at the garbage dump in Pires do Rio (GO), we can see a large amount of sophisticated packaging, garbage bags, cleaning materials, food waste with frozen product packaging, etc. We therefore need to know where these products are going.

Bringing up the problem was a very challenging proposal for us. This is because we have to discuss environmental and human aggression. In order to organize the discussion in a didactic way, the work will be divided into three chapters.

For the construction and development of this research, we sought support in the works of SCARL ATO 1992, BRANCO 1991, BARBIERI 1997, ALLEGRETTI 2001,

BORGES 1991, SIQUEIRA 2006, PAES 1991, among others, as well as articles researched on the internet.

In the first chapter, in order to lay the foundations for our work, we discuss the concepts of waste, garbage, landfills and sustainability.

The chapter begins with the concepts of garbage and landfill, looking at literature on the subject for clarity. Also in this chapter, we present the concepts of landfill, also supported by authors on the subject.

Finally, we are going to work on the concepts of sustainability. This topic is necessary because in the course of the research we will present our main focus, which is the Urban Solid Waste collectors.

In the second chapter, we get to know the city of Pires do Rio (GO), we also talk about the collection and final disposal of waste, and at the end we introduce the waste pickers, talking about their history and concepts of Urban Solid Waste Waste Pickers.

In this chapter we try to introduce our readers to the municipality of Pires do Rio (GO), from its founding in November 1922 to the present day, highlighting the main milestones in our history, which required reading the works of Pires do Rio writers. Next, we'll take a tour of the collection and final disposal of garbage at the municipal dump.

To close the chapter, we worked on the history and concepts of the waste pickers, who are the real soldiers in the fight for the survival of the environment, looking for a source of income from waste. They are men, women, young people, adults - honest people who have been excluded from the demanding job market and have found a sustainable source of income in scavenging.

In the third and final chapter, we will get to know the waste pickers at the municipal dump in Pires do Rio (GO). It was necessary to use formal conversations and interviews in order to get to know them better. We also talked to the municipal environmental department and to residents living near the dump.

At the end of our work, we present our final thoughts with a suggestion for the problem.

1 OUR DAILY GARBAGE

In order to gain a better understanding of the concepts of garbage and landfill, as well as the concepts of sustainability, and to provide a basis for the work that is to be presented in this research, it was necessary to analyze the concepts of garbage and landfill.

1.1 Concepts of waste and landfill

At the dawn of time, people were nomads. When food became scarce, they moved to another region and their "waste", left in the environment, was soon decomposed by the action of time.

At a certain point in time, people began to stay permanently in one place, producing their own food there, which led to an increase in the production of waste, which at the moment is not a global problem.

With the increase in population and the advent of the Industrial Revolution, new consumer habits and waste, waste piles up. The increasingly urban and sophisticated human being generates an even greater amount of waste, without worrying about the consequences.

Scientific technical advances and the growing industrialization process have progressively interfered with and altered the environment for the benefit of capital. In this process, some countries have seen the greatest changes, especially the developed ones (ROSS, 2003)

The concentration of population in cities and the huge increase in the consumption of goods generate a huge amount of waste of all kinds, coming from homes, public activities and industrial processes. All these materials are called waste, a great diversity of solid waste from different sources, including the urban solid waste generated in our homes.

The inhabitants of modern society, especially those in large urban concentrations, have at their disposal a very wide range of consumer goods. Ultimately, these are pieces of relatively complex material, both from the point of view of their structure and their energy content (SCARLATO, 1992, p.51).

The rate of generation of solid urban waste is related to the consumption habits of each culture, where there is a close correlation between waste production and the economic power of a given population. Waste is part of human history, since its production is inevitable, and its elimination and possible reuse are challenges still to be overcome by modern societies.

Solid waste is just one of the problems degrading the environment, and its nuisance is caused by the lack of awareness on the part of the population, who throw rubbish in inappropriate places. Nature can no longer keep up with the production of garbage and so the damage it suffers is increasing (BRANCO, 1991, p. 50).

Waste, as it is commonly called, is any residue that is surplus to the processes of industrialization and human consumption, and it is being discussed frequently. This is the price we pay for our industrialization, our consumerism and our daily comfort. However, it is believed that the problem is not the amount of waste produced, but what is not done with it - proper treatment.

Everything that is thrown on the ground not only makes the ground filthy, but also annoys people wherever they go and harms the environment, which in most cases is unable to decompose, as is the case with rubber. A simple piece of paper can take up to six months to decompose, while glass takes a million years to disappear from the environment.

Table 1 - Solid waste decomposition time

Garbage	Decomposition time
Fruit peels	1 to 3 months
Paper	03 to 06 months
Cloth	from 6 months to 1 year
Chewing gum	05 years

Cigarette filter	from 05 to 10 years old
Bottle cap	15 years
Painted wood	15 years
Nylon	more than 30 years
Plastic bags	30 to 40 years old
Tinned food	100 years
Aluminum cans	200 years
Plastic	450 years
Disposable diapers	600 years
Glass bottles	undetermined
Tire	undetermined
Plastic bottles (pet)	indefinite
Rubber	indefinite
Glass	1 million years

Source: 10th SILUBESA - Luso-Brazilian Symposium on Sanitary and Environmental Engineering. Org. Paulo Jean Pereira

The natural cycles of decomposition and recycling of matter can reuse human waste. However, a large proportion of this waste overloads the system. The problem is exacerbated because many man-made substances are not biodegradable, i.e. they don't decompose easily. Glass, cans and some plastics are not biodegradable and take many years to decompose. This waste can cause pollution

It was with this in mind that the systematic urban cleaning service was officially launched in Brazil on November 25, 1880, in the city of São Sebastiao do Rio de Janeiro, then the capital of the Empire. On that day, Emperor Pedro II signed Decree No. 3024, approving the contract for the "cleaning and irrigation" of the city, which was carried out by Aleixo Gary and, later, by Luciano Francisco Gary, from whose surname originates the word gari, which today is used to refer to urban cleaning workers in many Brazilian cities. (GUERRA, 2004, p. 58)

As urban waste management is essentially a municipal activity and the activities that make it up are restricted to the territory of the municipality, consortium solutions are not very common in Brazil, even when it comes to final disposal in landfills.

Society didn't care or didn't realize how much garbage it was producing and the harm it was doing to the environment, so it used any free area to dump its garbage. With this increase in waste produced, garbage dumps and landfills came into being.

The problem of final disposal is of alarming magnitude. Considering only urban and public waste, what can be seen is the widespread action of local public administrations over the years to simply move collected waste away from urban areas, sometimes depositing it in completely unsuitable locations. The waste is simply dumped in the open as shown in figure 1, without any sanitary criteria to protect the environment, which allows full access for disease vectors such as flies, mosquitoes, cockroaches and rats, such as forested hillsides, mangroves, rivers, bays and valleys. More than 80% of municipalities dispose of their waste in open air, in watercourses or in environmentally protected areas, most of them with the presence of waste pickers, including children, highlighting the social problems caused by poor waste management.

Figure 1 - Municipal dump of Pires do Rio (GO) Author: ALVES, Paulo Jean Pereira, 2009.

Thus, a dump is an inadequate way of disposing of solid waste, characterized by the simple dumping of waste on the ground, without measures to protect the environment or public health. It is the same as open dumping. In open dumps[1] , there is no control over the type of waste deposited and where it is disposed of. In these cases, low-hazard household and commercial waste is dumped together with industrial and hospital waste, which is highly polluting (INSTITUTO DE PESQUISAS TECNOLÓGICAS DO ESTADO DE SÂO PAULO, 1995).

It is the place where urban and industrial waste accumulates in a rustic way, without any treatment. As there is no treatment for the waste deposited in the dump, which is mixed with household, hospital and industrial waste, it causes various environmental problems, as well as the needy and unemployed population who are attracted to the dump, most of whom feed on the waste found there.

According to the Brazilian Institute of Geography and Statistics (IBGE), 40% of municipalities dispose of their garbage in dumps, which are institutionalized, i.e. authorized by the municipal authorities. The pollution caused by a garbage dump reaches several kilometers around it. The accumulation of garbage in the soil causes a series of problems not only for certain ecosystems, but also for society. Contamination of the soil, the water we drink and the air, because spontaneous burning is constant, is a problem that Brazil has been suffering from for a long time.

As the dumps are neither controlled nor measured by anyone, any uninformed person or company can dump dangerous waste there, such as hospital waste, radioactive or very toxic products, which should have special treatment.

1.2 Landfill

All over the world, with a few exceptions, landfills represent the main final destination for solid waste, despite the immense effort to reduce, reuse and recycle. Despite the contradiction, in developing countries, landfill has been the most important goal to

1 A term that can also be called a dump.

achieve in order to treat waste properly.

According to Gonçalves (2006), there are *already* a significant number of sanitary landfills in Brazil, especially in the Southeast and South. On the other hand, in the other regions, these facilities are practically non-existent. The great difficulty lies in the cost of operating a landfill, which presupposes adequate treatment of effluent liquids and gases.

A landfill is a space for the final disposal of solid waste generated by human activity. It is used to dispose of domestic, commercial, health service, construction industry and sewage waste.

The Brazilian Association of Technical Standards (ABNT) defines sanitary landfills as follows: Sanitary landfills of solid urban waste consist of the technique of disposing of solid urban waste in the soil, without causing damage or risks to public health and safety, minimizing environmental impacts, a method that uses engineering principles to confine solid waste to the smallest permissible volume, covering it with a layer of earth at the end of each working day or at shorter intervals if necessary.

Scarlato (1992) suggests that landfills are places where waste is confined without causing major damage to the environment. It is a method in which the waste is compressed by means of machines that reduce its volume. With the work of the tractor, the waste is pushed, spread and kneaded on the ground (compaction) and then covered with a layer of sand, which minimizes odours, prevents fires and stops the proliferation of rodent insects.

Most of the time, they are built far away from cities. This is because of the smell and the possibility of soil and groundwater contamination. In Brazil, the minimum distance from a landfill site to a watercourse is recommended to be 400m. They must have control over the quantity and type of waste, environmental protection systems and environmental monitoring. Landfills are important because they solve part of the problems caused by the excess waste generated in large cities. (SILVA, electronic doc.)

Ideally, it should have a useful life of more than 10 years, and it should also be monitored for a few years after closure. In the process of decomposing solid waste, highly polluting gases and liquids (leachate or percolate) are released, which means that a landfill project requires care such as waterproofing the soil, setting up effective drainage systems, among others, to avoid possible contamination of water, soil and air.

Once leachate is formed, it is a liquid substance resulting from the process of putrefaction (rotting) of organic matter. It is viscous and dark and has a very strong, unpleasant smell. It must be drained and taken to a treatment system before being discharged into a body of water. In Brazil, when it comes to treating leachate *in situ*, biological lagoons are often used, which have the difficulty of requiring a very large area in regions with high rainfall and humidity levels (BRANCO, 1991, p. 78).

The process of treating leachate is very important for the environment. If it is not treated, it can reach groundwater, rivers and streams, leading to contamination of these water resources. In this case, fish can be contaminated and, if the water is used for agricultural irrigation, the contamination can reach foodstuffs (fruit, vegetables, etc.).

In this sense, many projects use the leachate recirculation technique to reduce the amount of liquid to be treated, but in rainy seasons the system can reach the limit of its capacity. Leachate recirculation should be applied when the humidity or degree of saturation of the waste is monitored, as it not only increases its specific weight, but can also inhibit the biodegradation process. On the other hand, the biological lagoon treatment system does not reduce the amount of nitrites, nitrates and ammonia present in the environment, which can lead to contamination of watercourses. Studies are being carried out in order to define new technologies for efficient and cost-effective leachate treatment.

Landfill gas treatment is practically all done by burning methane (CH_4) and releasing carbon dioxide (CO_2). In general, the gas drainage system is individual (open type) using flare burners[2] , with some exceptions in combined drainage systems with forced gas extraction. Gas treatment with an open type burner is not the most advisable since the burning is not complete, as is the case with enclosed type burners. Incomplete or low-temperature combustion not only does not guarantee the transformation of

2 A flare is a burner fitted with a continuously lit pilot to ensure that the combustible gas burns in the open air and in a safe place.

methane (CH_4) into carbon dioxide (CO_2), but can also result in the release of toxic compounds that could be destroyed by the action of high temperatures.

1. 3 Sustainability

We see on a daily basis how much human beings are destroying the environment. The growth of cities, industries and vehicles are causing damage to the air, soil and water. Development is necessary, but human beings need to respect the environment, because we depend on it to survive on this planet.

Sustainability is a complex subject and difficult to measure. It aims to be a means of shaping human civilization and activity in such a way that society, its members and its economies can fulfill their needs and express their greatest potential in the present, while at the same time preserving biodiversity and natural ecosystems, planning and acting in such a way as to achieve pro-efficiency in the indefinite maintenance of these ideals.

Sustainability is the quality of what is sustainable and comes to incorporate the meaning of maintaining and conserving natural resources. This requires technological advances that permanently apply the capacity to use, recover and conserve these resources, as well as new concepts of human needs to relieve society's pressures on them (BARBIERI, 1997).

> Sustainable development is development that meets the needs of the present without compromising the ability of future generations to meet their own needs. This development also presupposes a commitment that tells us that it is impossible to live with hunger, poverty and misery today, and another that makes us aware of the need to ensure the means of survival for future generations (ALLEGRETTI, 2001, p. 54).

Sustainable development means achieving the necessary economic development while guaranteeing ecological balance. It encompasses various levels of organization, from the local neighbourhood to the entire planet, with continuity in the economic, social, cultural and environmental aspects of human society.

> Sustainable economic development, in other words, is one in which real income grows as a result of the growth

of the economy's real productive factors and not in nominal terms, which would be unsustainable growth because money would just be thrown into the economy, generating momentary wealth which, when economic agents notice that there is no equivalent production to this artificial income gain, they adjust their prices, which in turn causes inflation. (JACOBI, electronic doc.).

This idea came about to find new ways of controlling or sustainably defining the destination of all waste, especially those most responsible for environmental pollution.

The concern with sustainable development represents the possibility of guaranteeing socio-political changes that do not compromise the ecological and social systems that sustain communities.

The growth of cities means that the environment takes on a new configuration, creating environments specific to the urban fabric. The increasing growth of cities or urban-industrial society has been consolidated with consumerism as an ideology of life, increasing the volume of domestic and industrial waste in both advanced and underdeveloped societies.

The introduction of new technologies into production processes in many countries, with the aim of raising production levels and reducing costs, is a compliment to the technical and rational knowledge peculiar to capitalist society. These, in turn, are appropriated by the owners of the means of production, who use them to improve the quality of their products and reduce their costs in order to compete in the market.

Aware of the need to understand more about the history of the city of Pires do Rio, about the collection and final disposal of solid urban waste and about the people who make a living from garbage collection, this monograph will turn to this next chapter.

2 GARBAGE AS A MEANS OF SURVIVAL IN THE CITY OF PIRES DO RIO (GO)

In the second chapter, we will present a brief history of the city of Pires do Rio (GO), then a brief exposition of the collection and final disposal of Municipal Solid Waste and finally a little about the history and concepts of "recyclable material collectors".

2.1 A bit of the History of the City of Pires do Rio (GO)

The town of Pires do Rio was founded on November 9, 1922, with the inauguration of the Goiás Railway, shortly before the town was founded by engineer Balduino Ernesto de Almeida.

> With the construction of the bridge over the Corumbà River, the continuation of the line and the construction of the Pires do Rio station, which soon became a town, Roncador lost its economic significance and commercial hegemony. Most of the economic activities fell into decay or were transferred to the new towns and villages that sprang up with the extension of the line. It was the first city in the Midwest to have been previously planned (BORGES, 1991, P. 108).

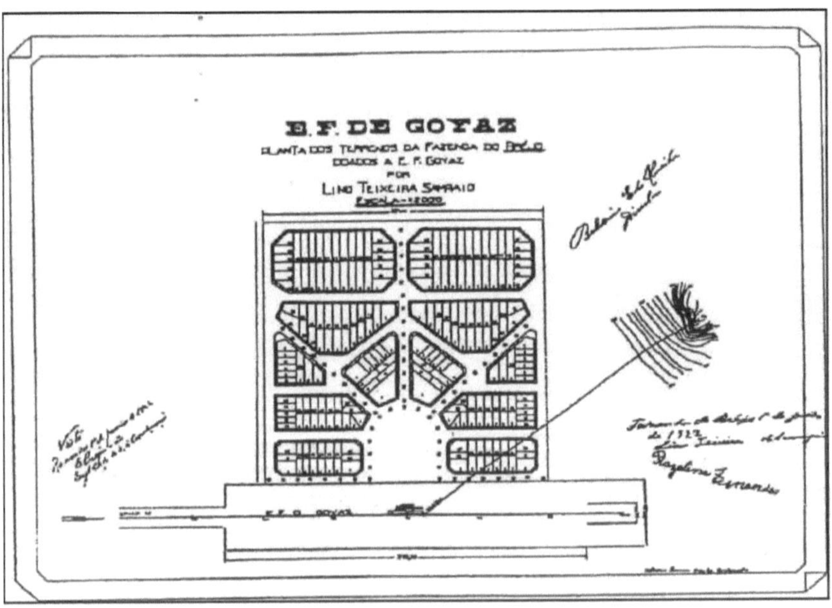

Figure 2 - Pires do Rio (GO), Plan of the municipality's urban area in 1920 Source: Railway Museum

The first plan of the city of Pires do Rio was not executed, but the second plan made by the topographer of the Moacir de Camargo Railroad, more in line with the topography of the place, made Pires do Rio the first city in the Midwest to be born with prior planning, on paper.

In 1920, he conceived the idea of installing a Railway Station, through the donation of four bushels of land by Colonel Lino Teixeira Sampaio, because with the construction of the Epitàcio Pessoa Bridge, he could continue the railroad. In 1922, the 120-meter-long Epitàcio Pessoa metal bridge over the Corumbà River was inaugurated.

The Sampaio farm (a group of farms: Brejo, Sumidouro, Marrecos and eight others belonging to Colonel Sampaio) was known as a landing place for drovers, due to its strategic location, and received illustrious guests who came to Rio de Janeiro and Sao Paulo, such as Sen.

Antonio Ramos Caiado (Totó Caiado), who always stayed at Col. Sampaio's house when he came and went to the Federal Capital.

For all the above reasons, no matter what they say, have said or written, the inference and possible conclusion is one: the founder of Pires do Rio, the only and exclusive one, by right and justice, is LINO TEIXEIRA SAMPAIO, excluding any other, due to the fact that, in this case, there is no origin or authority, but only the dream and the human will and this belonged to the old Santa-Cruzano (SIQUEIRA, 2006, p.38).

The origin of the name is a tribute to the Minister of Aviation, José Pires do Rio, who at the time came to inspect the construction of the Epitàcio Pessoa Bridge. First the name of the station, then the village, then the municipality.

The railroad arrived in Goiás around the 1920s, bringing improvements to the state. The locomotives transported passengers, making it possible to connect inland towns with other locations; they turned rural workers into train drivers and boosted trade. The Goiás Railroad transported the construction material that helped build Brasilia (PAES, 1991, p. 33).

A population center began to emerge around the station, which was designed by Alvaro Sérgio Pacca, chief designer of the Goiás Railroad. On August 28, 1924, Pires do Rio was elevated to a district by Municipal Law No. 66 of the Santa Cruz de Goiàs City Council.

The first neighborhood to emerge in the city of Pires do Rio was Bairro Santa Cecilia or commonly known as "Rua do Fogo". According to local residents, the name "Rua do Fogo" came about because it was the first street in the city, and it was along it that the muleteers passed, and during the night they would light a fire to cook food and warm themselves up during the early hours of the morning.

In 1930, Pires do Rio was elevated to city status. The municipality was created by State Law No. 903 of July 7, 1930. The first residents of the town were farmers who were already living there when it was founded.

> The village of Sampaio grew with the arrival of families from Santa Cruz de Goiàs and the urban agglomeration around the Roncador Station, which for about nine years was a line point, with flourishing commerce, pensions, commission houses and a cinema. With the growth of the town and the railway station bearing the name Pires do Rio on both sides, it became the new town. (SOARES, 1995 p.20)

Later, people came from other cities and neighboring states, mainly Minas Gerais, and from other countries, such as the Syrian families who settled here as traders.

Pyrrhic society was divided into two parts: the "elite" and the "proletariat", the former represented by the colonels and their families and the latter by the settlers and servants.

Then came the middle class made up of small landowners and small traders. Today the social division in the city can be seen on the basis of economic power.

Pires do Rio is 142 km from the state capital, Goiânia, located in the Mesoregion of Sul Goiano, Microregion of Pires do Rio, in the state of Goiás, in an area of 1,073.369 square kilometers (Figure 3).

It is served by the GO-020 and 330 highways, as well as several municipal roads. Its relief is characterized by plateaus, hills, plateaus and small undulations.

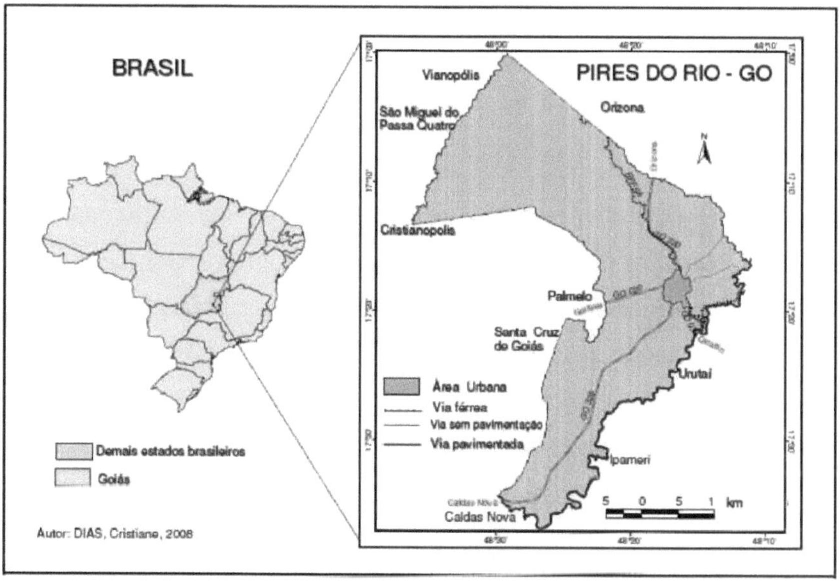

Figure 3 - Municipality of Pires do Rio (GO) Source: DIAS, Cristiane, 2008

Figure 3 shows the privileged location of Pires do Rio, which has made it a hub for the surrounding towns. People from the surrounding towns come here to do their shopping of various kinds, and also to get medical care.

The city of Pires do Rio (GO) currently has approximately 30,000 inhabitants, who enjoy a complete water supply, energy and sewage treatment service. In the health sector, the city has private clinics, a municipal hospital and first aid centers in the neighborhoods (Psfs). In education, there are municipal, state, private and public schools, as well as two higher education institutions.

2.2 Collection and final disposal of solid waste in the city of Pires do Rio (GO)

As has already been discussed, at first the town of Pires do Rio (GO) was a small village where the inhabitants produced little waste, which was thrown into backyards and vacant lots where most of it was turned into organic fertilizer. The arraial then became a city, and industries or industrialized products began to arrive, and with this the daily production of waste increased, so around the year 1950, with a population of approximately 5,000 inhabitants, the daily collection of waste by the municipal government began.

Há at first, collection was done by animal traction, in a simple way. As the population grew, there was a need to improve the collection of solid urban waste. Tractors and trucks were purchased and staff hired to carry out the collection. Waste is currently collected by an outsourced company, which collects it in the morning, afternoon and evening, in trucks equipped with a press system, which compresses the waste inside the bucket, thus giving rise to a new load of waste, hospital waste is collected in separate trucks, with a special system to receive the waste and taken to the municipal dump, according to figure 4.

Figure 4 - Ditch for hospital waste at the municipal dump in Pires do Rio (GO) Author: ALVES, Paulo Jean Pereira, 2008.

The city of Pires do Rio has already had six (6) final waste disposal areas, the dumps, the last of which is still in use today.

Initially, the waste was collected and deposited on plots of land near the town, until the current dump was built. This site was designed and built to be a recycling plant, but so far it has been nothing more than an open-air dump. The town dump is located 2.2 km northeast of the town's urban perimeter, covering an area of 4.8 hectares (see annex 4).

The municipal government has been working satisfactorily in terms of cleaning up the city, seeking the support of society to keep the city clean, wooded, landscaped, etc. Recently, a project entitled: Aterro de residuos sólidos urbanos (Urban Solid Waste Landfill) was approved, with the aim of recovering the area of the dump and preparing it to receive the project.

The area *has already* been fenced off, and receives large quantities of garbage and urban debris every day, without any planning, being an area with no cut-outs or roads, for the passage of dump vehicles and tractors that work without any criteria in the process of compacting the waste. In addition to these factors, there is the presence of waste pickers, who daily collect, sort and sell materials that for many are just garbage, but for them are a source of income.

2.3 Waste Pickers - History and Concept

The "waste picker" profession has existed since the mid-1950s. Waste pickers have always been seen as socially excluded. However, waste pickers have always provided a service to society, even without receiving the recognition or payment they deserve.

Figure 5 - Urban garbage collector

Source: folhavipdecaiazeiras.blogspot.com/2008

Collecting garbage is an alternative found by some people who are excluded from modern society, even if they are old or even young, as in the case of figure 5, since they don't have the qualifications required by the market, they see this job as a survival strategy. Even though this form of work is seen as degrading by society, the "waste pickers" have made garbage a way of earning an income to support themselves.

The question of survival and the lack of options is one of the strong reasons why these people have taken up this profession, since due to their lack of qualifications they are not accepted by the job market and end up being excluded by society. They are considered to be self-employed, but if you make a comparison between before they became waste pickers and today, they mostly prefer today's work. It may not be a job that is well regarded by society in general, but it is bringing dignity to many of them.

In the past, as today, many waste pickers work precariously, in rubbish dumps and unsuitable places. Many still suffer humiliation and exploitation by scrap metal dealers and large recycling companies.

According to Carvalho (2006), large industries produce products that enrich only the ruling class, which in turn exploits its employees. This industry puts its products on the market, makes a profit, but is not responsible for the packaging and waste it produces. All the waste disposed of ends up in landfills or dumps. The waste picker, excluded from the production process, survives on what industry and commerce reject.

They are historical players in waste management in cities and in the recycling production chain and deserve public policies that strengthen their entrepreneurial and ecological profile.

> There are practically two types of waste pickers that everyone knows exist, but society simply makes a point of ignoring: those who collect waste directly from the dumps, in all cases with a view to selling it. (PORTUGAL, Electronic doc.).

Waste pickers have historically made a living from waste. Either through the practice of selective collection with some partners who donate their waste or, better still, their recyclables selected at source; or by hunting for recyclables in the streets and dumps, pulling recyclables out of the mixed waste that the generator hasn't had the decency to separate and putting in the same bag what can and can't be reused.

In this third and final chapter, more will be said about the waste pickers and their daily routine, and it is also necessary to get to know the waste pickers at the Pires do Rio dump and their survival.

3 COLLECTION AND VIABILITY FOR THE DUMP SURVIVORS

In this third chapter, we will learn more about urban solid waste collectors. To do this, we'll need to interview the waste pickers and visit the municipal dump. We'll also check with the government to find out what actions the municipality is taking with the waste pickers and, finally, we'll talk to some residents of the area near the dump to find out their ideas about the waste pickers.

3.1 Valuing waste pickers

For more than half a century, Brazil has been home to waste pickers, a silent class of workers of great importance to society, the environment and the economy.

They suffer disrespect every day, from physical aggression to assaults in the streets, because part of the population considers the way they present themselves to be marginalized, giving the city as a whole a disorganized look. However, despite all the lack of encouragement, this class has been trying to occupy a space of respect in society, aiming to organize themselves in the form of cooperatives and associations.

This professional class plays a fundamental role in society as a whole, preserving natural resources, reducing solid waste destined for the dumps, as well as interfering in social issues, as they are people who are excluded from the demanding job market and find an employment opportunity in recycling.

The "garbage collectors" are looking for ways to survive other than those set by the technical-productive system, which does not mean that they are being autonomous in their survival process. They only seek this alternative because they can't fit into the patterns set by the system. And when they take to the streets to "collect" garbage, they don't do so out of a new awareness, with a desire for change. They take to the streets, working in this activity, in order, in a way, to remain included in the system, in order to acquire the consumer goods it provides.

So much more is needed than alternative practices. We need changes in thinking, awareness, culture and social change so that we can recreate our customs and habits in favor of the existence of human life, outside of this system that enslaves those who are part of it and marginalizes those who don't fit its standards.

As far as the environment is concerned, it is obvious that the activity of the waste pickers has brought benefits to nature and to people's attitude towards the culture of separating what is recyclable from what is waste. Since they themselves already separate the waste from their homes, we can see a great advance in this society when, not so long ago, hardly anyone paid attention to what could degrade the soil and nature. And as these "agents" work on the streets, they end up encouraging people to separate their waste. Over time, this habit will certainly become part of people's daily lives.

Collecting recyclable materials is, for many workers, the only way to ensure survival and the possibility of inclusion in an exclusionary labor market. Workers are exposed to dangerous conditions, are victims of prejudice and stigma, and are excluded from some social environments. They often work for more than twelve hours non-stop, an exhausting job given the conditions to which they are subjected, with their carts pulled by human traction, carrying more than 200 kilos of garbage a day, an average of four tons a month, and covering more than twenty kilometers a day.

In the end, they are often exploited by the owners of the garbage dumps who, in a paternalistic gesture, exchange the day's collected waste for alcohol or pay them a symbolic amount that is insufficient for their own reproduction as garbage collectors.

These scrap dealers receive the material collected by the waste pickers, weigh it and set the price to be paid to the waste pickers. In their warehouses, they accumulate the materials by pressing them into bales until they reach a quantity that allows them to be transported to the recycling industries.

The dump has become the scene of scenes that show the extent to which social exclusion promotes human degradation. Every day, these people go to the dump in search of useful objects or food. In some places, waste pickers are supported by government bodies or NGOs, because this activity helps the environment and also the individuals who sell the waste to ensure their survival.

This support for the organization of this class led to the founding of the National Movement of Waste Pickers (MNCR). This movement was born out of the encouragement of people linked to the Catholic Church, who, sensitized by the precarious life/work conditions of the waste pickers, encouraged their mobilization as a way of improving the conditions for this group to enter the recycling market. Thus, the formation of the MNCR has its origins in the work carried out with homeless people in some of Brazil's state capitals, most of whom are waste pickers.

Once organized in the MNCR, the workers are seeking, in addition to better working/life conditions, recognition from the society that marginalizes them and discriminates against the activity they carry out (waste picking). Without this mobilization, the workers would remain isolated and invisible, without the possibility of presenting themselves as the subjects of a broad process of change that transforms not only the conditions in which the work of scavenging is carried out, but the very social and historical process that has led millions of workers to survive miserably.

Thanks to the efforts of the MNCR, today waste pickers are being trained all over Brazil, through courses, congresses, meetings, among others, which is undoubtedly a recycling of their history.

Figure 6 - MNCR symbol

Author: National Movement of Waste Pickers.

In November 2002, united through the MNCR, which represents the country's waste pickers, they won a major victory by having their occupation recognized by the Ministry of Labor. Although it doesn't represent the end of the waste pickers' struggle, this achievement was worthwhile as an incentive to strengthen the union in the search for their rights and their dignity, which even though they are often treated as beggars or potential bandits, they prevent the felling of several trees a month, for example.

Waste pickers are responsible for collecting 200 tons of office paper every month in the largest Brazilian capitals. In addition to paper, other components of the waste produced daily in the city, such as glass, aluminum cans and PET bottles, are recycled thanks to the work of the waste pickers.

3.2 Solid Waste Collectors in Pires do Rio (GO)

In the city of Pires do Rio (GO) we find urban waste pickers and the municipal dump. The waste pickers who work in the city collect waste such as cans, office paper, cardboard, pet waste, among others, and sell it to companies in other locations.

It's common to find can collectors at parties and in public places. The cans are collected, crushed and sold to scrap dealers, who sell them in larger quantities to recycling companies. We can say that these can collectors are collaborators with public cleaning, because during or after a party or even a fair[3] their presence is common, according to figure 7.

Figure 7 - Can collector after a night of carnival in the town of Pires do Rio (GO) Author: ALVES, Paulo Jean Pereira, 2009.

They collect all the Iatinhas they can find, which used to be left behind and caused a frightening sight the next morning. They crush them at home and sell them to scrapyards or recycling companies.

The waste pickers who work at the dump in Pires do Rio (GO) are forced to sort their waste in the same area as the dump due to the lack of selective waste collection. There they sort out what can be reused and transport it to the plant, where it is sorted again and then pressed. They sell to the city of Catalao and press cardboard, pet bottles, soft

3 Local handicraft fair, held every two weeks in the central square of Pires do Rio (GO).

plastic, aluminum and scrap metal. There are an average of fifteen waste pickers, almost all from the same family.

Their daily routine doesn't vary much, as they wait for the urban collection trucks to arrive and start collecting, where they go through the garbage in search of materials, sharing spaces with insects, animals, among others. They stay in the dump area for an average of ten (10) hours a day. The garbage is collected and taken to the plant, where it is sorted and separated for later pressing.

In a conversation with Mrs. Ivana, a waste picker and representative of the other waste pickers at the municipal dump, she told us a little about her story. A little over ten years ago, she and the other workers at the dump were urban workers, from where they made their living, but Mrs. Ivana, in a conversation with her husband, decided to go and work at the dump, and this decision aroused a lot of prejudice, because as we know, but she and her husband didn't let themselves be defeated, they went to the dump and soon after some of her family, including her son, brothers and sisters-in-law, who are still working in the dump to this day. She also told us that they use almost everything from the garbage, but the best trade is cardboard, pet bottles and some aluminum.

As has already been mentioned, waste pickers are forced to share their collection with agents that are harmful to their health, either invisible vectors made up of bacteria, or animals that live there, such as rats, flies, cockroaches and dogs. They work with little or no infrastructure.

In a conversation with another waste picker who asked not to be identified, he commented that: "When we're collecting in the dumpsite area, we use some protective equipment, including gloves, long-sleeved pants and shirts, and boots. We prefer boots, because rubber boots are less durable and get punctured a lot. We know how important protective equipment is, but since we don't receive much help from politicians, we don't have enough money to buy it.

After the waste has been collected in the dump area, it is taken to the plant, where it is sorted and then pressed. According to Ivana, the town hall has given over the area of the plant, with the right to use both the building and the equipment, and the town hall also provides the energy. Once the waste has been selected, it is pressed and stored for sale to a company in Catalao (GO) when there is a good quantity (see figure 8). Our waste picker told us: "At other times the prices were better, but today it's not so rewarding, but I still prefer, both me and my companions, here at the dump than having to work as maids or be Nutriza employees[4] in the city".

When we talked about her approximate income, Ivana was satisfied with the amount, she didn't make it clear how much, but she told us that it's enough to live well. She said: "It's difficult to give you a real figure, but on average I earn between one and two times a month, but of course it varies a lot, some months the company pays better, other months they pay poorly".

Figure 8 - Pet bottles and pressed cardboard ready for sale Author: ALVES, Paulo Jean Pereira, 2009

Dona Ivana tells us about her profession, her eyes full of tears: "it was like I told you

4 Nutriza Agroindustrial de Alimentos S.A. - Complex made up of poultry feed factories, breeder farms, hatcheries and a chicken slaughter and industrialized products processing unit.

about my decision to come and work here, I was in the city with a lot of bills due and many to come, and my partner was out of work too, we decided to try our luck here, we managed to turn things around, I cleared my name and that of my family, I feel good here, because this service, Today I have my savings, I have a little motorcycle that takes me to various places, I go to the city when I want without having to hitch a ride with the garbage trucks, it's like I told you, I wouldn't trade this job for one in the city, even when prices are low".

As mentioned in chapter two (2), urban waste collection is the responsibility of an outsourced company, and talking to our waste pickers, they told us: "the organization of the dump has improved a lot, because before, the city hall trucks would arrive and dump the garbage anywhere, often without caring even for those who were there, which made it very difficult to collect, and not to mention that most of the time the garis themselves, collected the cans and some aluminium, which made us angry with them, today this no longer happens, the truck arrives and unloads the garbage in a place that is still vacant, and the garis don't keep collecting the cans or any garbage that will help us earn a little money, only one thing that is still wrong is that even though the company is more organized than the city hall, when the dump is full the city hall sends a tractor to compact the garbage, and it does it anyway, which makes our job more difficult".

At the end of my conversation with Mrs. Ivana, I asked her what her hopes were for the future, and she said: "My son, I think a lot about everything I've done, and I don't have many hopes left, I just want to keep my name clean, have friends and I'd also like to be seen as a person and not as an animal, that's all, if I can achieve that I'll die in peace".

In a conversation with Mr. Luis Mauro, Technical Director for the Environment, he told us that: "The Municipal Environment Department has no direct relationship with the dump collectors, because even though they have been there for several years, whenever a meeting is needed to try to get them out of the site, they resist leaving and often don't show up for the meetings."

When I asked him if he knew any waste pickers, and what he thought of the waste picking profession, he replied: "Yes, I do, I think it's a very degrading profession, living in the midst of so much garbage, with no health conditions, hygiene, etc. I think they should have a fixed salary, because they only earn what they can collect from the landfill, they need a cooperative. I see it as a dignified profession, but at the same time very degrading."

In an interview with Dona Silvia, a resident of the Vila Mutirao neighborhood, located at the exit to the garbage dump, she said: "I admire the pickers, I don't have friends with any of them, but they are normal workers, if I had better health, I would be a picker too, I know by sight some who work there, I think they work a lot, but it was like I told you if it was another time, I was there too".

Speaking to a young man, who asked not to be identified, he told us: "these people who work at the dump are really excluded, because there are jobs in the city, you don't need to be in a place like that, in the middle of so many animals, you have to be very lazy not to take on heavy work, to be in a place like that, I had a schoolmate who was a picker at this dump, but he was ashamed to tell what he worked with, I think even these people who work with garbage, see their profession as shameful"

Talking to Dona Antonia, the owner of a supermarket on the way out to the dump, she tells us: "I always see these people who work at the dump passing here on the avenue, sometimes they stop here at the Supermarket to buy something, I've never been to the dump, but I imagine it's a frightening sight, I think it's a stressful profession, sometimes embarrassing, but at the same time it's a situation of pride for them, because my idea is that I think it's better to work, even if this service is different from the others, it's better than stealing. But I see this profession as very hard, and not worthy of anyone.

After talking to the waste pickers, it became clear that they see their earnings from working at the dump as superior to those received in their previous profession. Another benefit, according to them, is their autonomy in managing their time, as they can work as many hours a day as they want and on the day they want.

Through *on-site* observation, we were able to see that waste pickers are exposed to harmful environmental agents commonly through inhalation, dermal contact, food contamination, as well as occupational risks to which they are subject, such as various accidents, cuts, being run over by trucks, among others.

One of the greatest dangers in this occupation was the possibility of cutting their skin with sharp materials, such as shards of glass, blades and wood splinters found among the debris.

The apparent utopia of an environment that reconciles sustainability, quality of life and social equality will only be achieved with a great deal of reflection, good will, personal and community effort, willingness and political action allied to the fundamental understanding that the planet as a whole is affected by each isolated attitude.

FINAL CONSIDERATIONS

At the end of this monograph, we realized that all the factors discussed in this study must necessarily be associated with each other in order to achieve the sustainability that is hoped for. The social factor, which is often overlooked, is very important because it deals with the human aspect, from the moment it includes in society individuals excluded from the job market and social life, as is the case with waste pickers.

It is of fundamental importance that these social actors have professional recognition, but above all, that they benefit from good working conditions, as this is a fundamental part of the process of sustainable formation concerning solid waste, collaborating with the sustainability of the environment, society and the economy.

It will also be possible to increase the value of recyclable materials in the labor market, thus creating a sustainable cycle.

The issue of sustainability is a multidisciplinary one, which means that it must involve various sectors if it is to work fully. Society needs to learn to respect and see waste pickers as workers and not as the marginalized individuals they are.

Even though they work in precarious situations, waste pickers are extremely profitable elements in the recycling production chain.

Today, waste picking is recognized as a profession, and this is an indication of change in the lives of these workers. We should also point out that there are organized groups fighting to improve their working conditions, including the National Movement of Waste Pickers (MNCR).

This raises the question of how to improve the working conditions of waste pickers. Organizing in cooperatives may be a way out, as they present a logical and innovative solution within the capitalist world, but as long as it is linked to both public power and private initiative. Organized in cooperatives, they provide an alternative way of strengthening their work, always focused on improving their work.

Once the cooperative has been legally created, it is important for the government to offer support in order to make up for the basic shortcomings that hinder the good performance of an organization, especially at the start of operations. We can highlight, among other actions, support for hiring staff with experience and training in administration and accounting; the provision of personal protective equipment and incentives for its use; the implementation of literacy and environmental education programs for waste pickers and their children.

And publicizing the importance of the work of these waste pickers is very important, because society would then get to know and respect them as decent, hard-working people.

REFERENCES

ABNT Associaçao Brasileira de Normas Técnicas NBR 10.004/04 - "Residuos Sólidos - Classificaçao. Rio de Janeiro: ABNT, 1987.

ALLEGRETTI, Alessandro, Explaining the Environment. 1st ed. Rio de Janeiro - RJ: LIS, 2001.

BARBIERI, José Carlos - Development and the environment: the strategies for change in Agenda 21 / José Carlos Barbieri. - Petrópolis, RJ: Vozes, 1997.

BORGES, B. G. O Despertar dos Dormentes, Goiânia: Cegraf, 1990.

BRANCO, Samuel Mergel, **Ecologia na Cidade**, 2° ed. Moderna, 1991. - (Collection

Challenges)

CARVALHO, Julia Maria Gomes e, Vida e Lixo: A Situaçao de fragilidade dos catadores de material reciclâvel em Marilia e os limites da reciclagem - 2006.

FINAL DISPOSAL OF SOLID WASTE IN BRAZIL: CURRENT SITUATION AND PERSPECTIVES - José Fernando Thomé Jucà - Federal University of Pernambuco, Brazil. *10º SILUBESA - Luso-Brazilian Symposium on Sanitary and Environmental Engineering.*

From Waste to Citizenship. Strategies for Action. National Waste and Citizenship Forum, 1998.

DREW, D. **Processos interativos homem x meio ambiente.** 4. ed. Rio de Janeiro: Bertrand Brasil, 1998.

DIAS, Cristiane. Mapping the municipality of Pires do Rio - GO: Using Geoprocessing techniques. Master's degree dissertation - Geography Postgraduate Program - Federal University of Uberlândia, 2008.

GONÇALVES, C.W.P. **The globalization of nature:** and the nature of globalization. Rio de Janeiro: Civilizaçào, 2006.

GUERRA, A. J.; CUNHA, S.B. **Impactos ambientais urbanos no Brasil.** Rio de Janeiro:

Bertrand, 2004.

GUIMARÂES, Mauro. Environmental education: Is consensus an embaté? - Campinas - SP: Papirus, 2000. - (Papirus Education Collection).

http://www.flickr.com/photos/hcordeiro/3903880617 (accessed September 2009).

http://www.ibge.gov.br/home/estatistica/populacao/condicaodevida/pnsb/lixo_coletado/defaultlixo.shtm - accessed April 26, 2009.

Brazilian Institute of Geography and Statistics (IBGE), National Survey of Basic Sanitation, PNSB, 1989, Sào Paulo.

INSTITUTE OF TECHNOLOGICAL RESEARCH OF THE STATE OF SAO PAULO (IPT). *Municipal waste*: integrated management manual. Sào Paulo: IPT/CEMPRE. 1995. 278p.

JACOBI, P. Educaçào Ambiental, Cidadania e Sustentabilidade available at (http://www.scielo.br/pdf/cp/n118/16834.pdf, accessed on November 18, 2008).

Jaqueline M° Dutra, Mariléia Alves da Silva. Solid Waste in the city of Pires do Rio - GO . TC (Graduation in Geography) - University Unit of Pires do Rio - State University of Goiás, 2003.

MENDONÇA, F. **Geografia e meio ambiente**. 5 ed. Sào Paulo: Contexto, 2003.

LIXO: desafios e compromissos, Pedro Sérgio Fadini and Almerinda Antonia B. Fadini, Revista Nova Escola, 2001.

SILVA, Marcelo, Projeto Lixo-Seletivo - Aterro sanitário available on the website (http://marcelosilva.com.br/proietolixo-03.htm) accessed on April 20, 2008.

PAES, Iranilda D. Rezende Pires do Rio, Nossa Terra Nossa Gente, Ed. Kelps, 1991.

Pires do Rio, Organic Law, Law No. VI Article 189, 190, 191 and 192.

PORTUGAL, G. O catador available on the website (http://www.gpca.com.br/gil/ant17 , accessed in August 2008).

PIRES DO RIO CITY HALL. Urban Solid Waste Landfill: recovery plan for degraded areas

- PRAD, 2005.

RODRIGUES, A. M. **Produçâo e consumo do e no espaço:** Problemàtica ambiental urbana. Hucitec: Sao Paulo, 1998.

ROSS, J.L.S. Industrial society and the environment. In: Geography of Brazil. Sao Paulo: Edusp, 2003.

SCARDATO, Francisco Capuano. 1939 - Do nicho ao lixo: ambiente, sociedade e educaçao / Francisco Scarlato, Joel Arnaldo Pontin; consultoria de Sérgio de Almeida Rodrigues, - Sao Paulo:Atual, 1992. - (Environment Series).

SIQUEIRA, Jacy. *Um contrato singular - e outros ensaios de Histórias de Goiàs*. Goiânia: Kelps, 2006.

SOARES, Francisco Accioli Martins. Historical Points of Pires do Rio - Ed. Piloto, (Nossa Folha edition), 1995.

VERNIR, Jacques, **O Meio Ambiente**. 2° ed. Campinas - SP: Papirus, 1994.

ANNEXES

Interview - Waste pickers

1. What do you take out of the trash?

2. Do you wear any protective equipment when collecting waste (boots, gloves, masks...)?

3. How much money does this help you with each month?

4. Are you satisfied with your work?

5. Has there been any change in garbage collection since urban collection was outsourced?

6. What is your outlook for the future?

Public Power Interview

1. How does the government relate to waste pickers?

2. Do you know anyone who lives off the dump?

3. What do you think about this profession?

Interview with the community

1. Do you know anyone who survives on garbage?

2. What do you think about this work?

MIX
Papier aus verantwortungsvollen Quellen
Paper from responsible sources
FSC® C105338

Printed by Books on Demand GmbH, Norderstedt / Germany